Computing and Systems Technology Division 2021

Held at the 2021 AIChE Spring Meeting and 17th Global Congress on Process Safety

Online
18 – 23 April 2021

ISBN: 978-1-7138-5340-4

Printed from e-media with permission by:

Curran Associates, Inc.
57 Morehouse Lane
Red Hook, NY 12571

Some format issues inherent in the e-media version may also appear in this print version.

Printed with permission by Curran Associates, Inc. (2022)

For permission requests, please contact AIChE
at the address below.

AIChE
120 Wall Street, FL 23
New York, NY 10005-4020

Phone: (800) 242-4363
Fax: (203) 775-5177

www.aiche.org

Additional copies of this publication are available from:

Curran Associates, Inc.
57 Morehouse Lane
Red Hook, NY 12571 USA
Phone: 845-758-0400
Fax: 845-758-2633
Email: curran@proceedings.com
Web: www.proceedings.com

TABLE OF CONTENTS

620660 The Digital Transformation Promises to Change Innovation in the Materials Industry. Will It?

Abstract:

The industry is challenged to innovate faster than ever while also delivering sustainability improvements that require profound change through the value chain. Simultaneously with that pressure, digitalization and AI are advertised as revolutions that has caused many to fear becoming obsolete. This dichotomy of extreme pressure and emerging, revolutionary technologies is causing fast, significant change in the industry and the innovation process. In this talk, I would like to share my view on what this impact will be, what risks exist, and what, in my view, will drive long-term success.

620661 Digital Twins & Smart Manufacturing –*Where Data Analytics Meet Modeling & Optimization*

Abstract:

Modeling and Optimization constitute core competencies of process systems and chemical engineering fields, and play pivotal role in the analysis, simulation, design and operational activities that we perform daily in order to improve the performance and sustainability of energy and manufacturing processes and supply chains. The re-emergence of AI and Big Data revolution that we are all witnessing at the moment, also provide an opportunity and a platform toward real-time decision-making. While often perceived as different fields, in the context of Smart Manufacturing, Data Analytics and Modeling/Optimization can in fact be viewed as 'two sides of the same coin'.

In this lecture, we will attempt to elucidate this argument through the lenses of ***hybrid modeling and digital twins***, which bring together the advantages and power of high-fidelity modeling & optimization and AI/machine learning technologies and tools, and their synergy. A central theme will be that the evolution/revolution in data-driven methods in the context of process modeling and automation is becoming a critical component in advancing the energy & manufacturing transition agenda toward clean energy, increased process/energy efficiency, sustainability and resilience, and ultimately economic and societal prosperity.

619569 Comparison of Statistical and Machine Learning Approaches for Fault Detection in a Continuous Catalytic Regeneration Unit

Abstract:

Catalytic reforming process is one of the most important processes in oil refineries that produce high octane number gasoline. However, its operation often leads to high coke formation on the catalyst particles, and subsequent degradation in reactivity. Continuous Catalytic Regeneration (CCR) system regenerates the catalyst continuously and helps maintain acceptable productivity goals. CCR accomplishes this by enabling the reformer to perform on lower coke ratio which is crucial for higher reformate yield. Any upset in the process operation directly affects the product formation and has adverse effects on economic yields. Hence, process monitoring using data-driven fault detection techniques is deemed to be crucial for ensuring reliable operating conditions.

Data-driven fault detection and diagnosis (FDD) methods, which can be divided into two categories, statistical and machine learning (ML) approaches, have become popular in both research and applications. In this study, we will present our key findings of process monitoring using both statistical and ML techniques. For the statistical approach, the commonly used method based on principal component analysis (PCA) will be employed and will be compared with dynamic PCA extension in which the serial dependence in measured variables is considered. For further comparison, on the same data, autoencoders, which is a popular unsupervised ML technique, will be employed for fault detection on real industrial operation data from the CCR process. This comparative study is expected to result in an effective fault detection approach for an operating plant that will also substantially contribute to process sustainability and reliability.

619061 Data-Driven Identification of an Industrial Reverse Osmosis Membrane Process: Impact of the Feed Properties, Plant Variables and the Operating Window

Abstract:

The petroleum facilities require large amount of water for the cooling, boiler feed, processing, sanitary services, fire protection, and many other miscellaneous purposes. Both feed water and resultant water in the refineries must be well treated due to corrosive effect, quality requirements and environmental issues. Treating the water via reverse osmosis (RO) membranes is an important industrial process in terms of sustainability ([1]). This study focuses on the data driven modeling of an industrial Reverse Osmosis Membrane Unit (ROMU) which includes cartridge filters, high pressure pump, and RO membrane series. RO membranes are composed of two stages with 36 and 18 membranes respectively. In more detail, one membrane module contains 6 spiral wound polyamide membranes as shown in Fig. 1.

<![if !vml]>

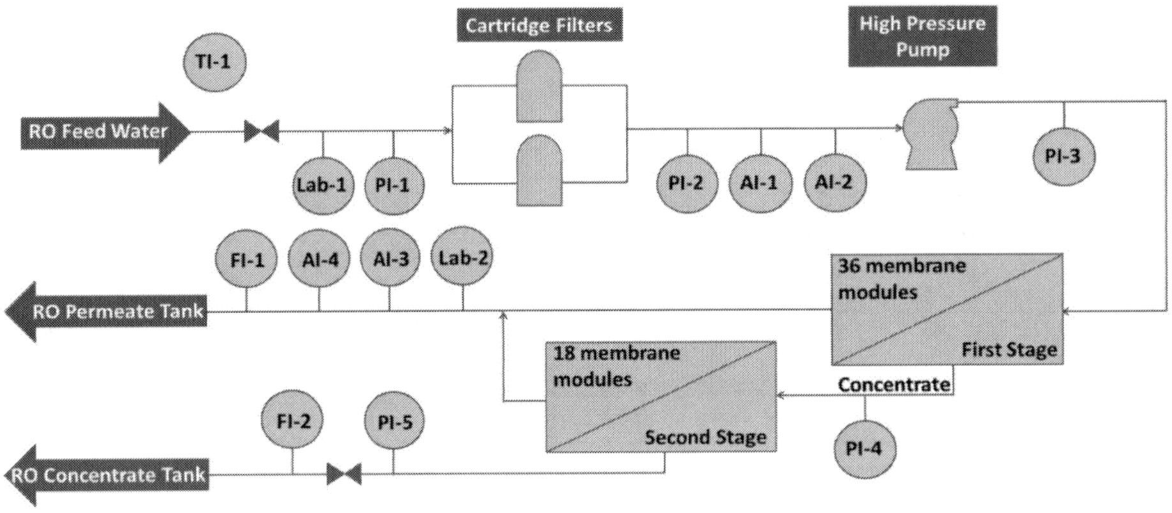

<![endif]>

Figure 1. Simplified ROMU diagram

Despite their wide usage, RO membranes suffer from permanent fouling once they are not maintained properly, resulting in higher cost and reduced product amount ([2]). In addition, the actual nature of the process is complex with high number of non-linearly coupled process variables. RO membrane fouling modeling and the determination of the major impacts on the process are the primary objectives of this study. Artificial Neural Network (ANN) is trained for the identification of the process to infer the impact of the plant and feed parameters on the outputs, primarily the product rate. The data for the model development contain the last 4 years, since the installation of the plant to date. The data include all sensors related to pressure, flow, temperature, conductivity, and pH. Next, the ANN is evaluated at probable operating conditions to calculate the operating window of the plant. The results show the impact of the process parameters on the product rate.

References

[1] Ismail, F., Khulbe, K. C., & Matsuura, T. (2018). *Reverse Osmosis*. Elsevier.

[2] Kang, G. D., & Cao, Y. M. (2012). Development of antifouling reverse osmosis membranes for water treatment: a review. *Water research*, *46*(3), 584-600.

619266 Elucidating Diafiltration Membrane Cascade Design Rules with Superstructure Optimization

Abstract:

The synthesis of many advanced materials in the past three decades have failed to produce a corresponding revolution in sustainable separation applications.[1] For example, new technologies for carbon capture can take up to 30 years to progress from a laboratory to commercial deployment.[2] To accelerate technology development, there is a critical need to move away from classical Edisonian iterative methods of materials discovery. Principled and data-driven frameworks will overcome the gaps in knowledge which inhibit the translation of new materials and devices from laboratory scale to sustainable separation technologies. In this talk, we describe our work using superstructure optimization at the systems scale and nonlinear parameter estimation at the materials scale in designing next generation membrane technologies and materials. We highlight how the current work lays the foundation for a holistic molecules-to-infrastructure computational design framework for the development of sustainable separation technologies.[3]

Relative to conventional technologies, membrane separations have demonstrated significant advantages in sustainability and energy efficiency.[4] For example, chemically selective separations enabled by polymer membranes have the potential to replace hazardous and toxic solvents used in conventional lithium-ion battery (LIB) recycling technology. However, there is no clear understanding of the interfacial and thermodynamic phenomena underlying the chemically-selective transport mechanisms needed to separate molecules with comparable sizes such as lithium (Li) and cobalt (Co) which hinders the development of membrane materials for this task.[5] Another challenge is the recovery of high purity products from a membrane separations unit. The formation of salt precipitate layers at the membrane interface when the feed side solution approaches the solubility limit prevents the extraction of high purity products, which also prevents the direct staging of these units.

In this talk, we demonstrate how a continuous, membrane based diafiltration cascade[6] enables the recovery of high purity Li and Co rich solutions for a LIB recycling case study at purities comparable with existing technologies. In diafiltration, the effects of salt precipitation encountered while staging membrane units are offset by adding a pure or dilute solvent, known as the dialysate to the feed side of the membrane filtration process. We show how the unique configuration of our system enables the utilization of the retentate side product, which is dilute in Li, to be recycled as dialysate to preceding stages of the process. Specifically, we highlight how superstructure optimization searches multiple, complex process configurations postulated by the modeler to identify promising applications for emerging materials. We treat flow and concentration of all streams as decision variables in our nonlinear optimization model. We then characterize the Pareto set using an epsilon-constrained method to understand the tradeoffs between purity and recovery objectives for the diafiltration cascade. Finally, we use the top-down capabilities of this framework to identify property targets for emerging membrane materials for LIB recycling.

Through our multi-objective superstructure optimization framework, we assemble a library of over 10,000 optimal designs spanning diverse system recovery specifications, area constraints, and membrane properties. From this library, we generalize common structures in process configuration into (near) optimal diafiltration cascade design rules. Many of these rules are non-trivial but ultimately justifiable by the governing transport and thermodynamic phenomena. In fact, some of these design rules show strong analogies between membrane diafiltration and distillation.

References

[1] Luo, J. & Crittenden, J. C. Nanomaterial Adsorbent Design: From Bench Scale Tests to Engineering Design. *Environmental Science & Technology* **53**, 10537 - 10538 (2019). 10.1021/acs.est.9b04371.

[2] Miller, D. C., Syamlal, M., Mebane, D. S., Storlie, C., Bhattacharyya, D., Sahinidis, N. V., Agarwal, D., Tong, C., Zitney, S. E., Sarkar, A., Sun, X., Sundaresan, S., Ryan, E., Engel, D. & Dale, C. Carbon Capture Simulation Initiative: A Case Study in Multiscale Modeling and New Challenges. *Annual Review of Chemical and Biomolecular Engineering* **5**, 301-323 (2014). 10.1146/annurev-chembioeng-060713-040321.

[3] Eugene, E. A., Phillip, W. A. & Dowling, A. W. Data science-enabled molecular-to-systems engineering for sustainable water treatment. Current Opinion in Chemical Engineering 26, 122-130 (2019). 10.1016/j.coche.2019.10.002.

[4] Phillip, W. A., & Elimelech, M. The Future of Seawater Desalination: Energy, Technology, and the Environment. *Science*, **333**, 6043, 712–717 (2011). 10.1126/science.1200488.

[5] Park HB, Kamcev J, Robeson LM, Elimelech M, Freeman BD. Maximizing the right stuff: the trade-off between membrane permeability and selectivity. *Science* **356**, 1-10 (2017). 10.1126/science.aab0530.

[6] Eugene, E., Phillip, W. & Dowling, A. Material Property Goals to Enable Continuous Diafiltration Membrane Cascades for Lithium-ion Battery Recycling. In *Computer Aided Chemical Engineering* **47**, 469-474 (2019). 10.1016/B978-0-12-818597-1.50075-8.

619247 Dynamic Modeling and Nonlinear Parameter Estimation of Nanofiltration Membranes

Abstract:

Compared to traditional chemical separation techniques, membrane processes offer compact configurations and energy-efficient means to achieve molecular scale separations. Membranes capable of achieving higher degrees of selectivity relies on both the precise control of a membrane's nanostructure and the identification of chemical functionalities which accentuate desired transport properties. [Hoffman & Phillip, 2020; Sadeghi et al., 2018] We argue process systems engineering offers untapped capabilities to balance membrane design decisions across (macro)molecular, device, and systems scales, and such a multiscale perspective is essential to fully realize the potential of nanoengineering to deliver transformative membrane separation technologies. [Eugene et al., 2019a] For example, we recently developed a superstructure optimization framework to design diafiltration membrane cascades. Our analysis shows that, through optimal systems-scale design, existing membrane materials have the potential to outperform current Li/Co separation methods for lithium-ion battery recycling; in other words, this particular application is mostly systems, not materials, limited. [Eugene et al., 2019b] In another study, we developed a process targeting framework to identify the necessary molecular and materials properties to realize absorptive membranes for environmental remediation. [Eugene et al., 2020] But extending these PSE paradigms to more directly inform nanoengineering requires more detailed and realistic mathematical models. [Geise et al, 2014; Yaroshchuk et al., 2018] Often the iterative creation and validation of said models is time-consuming and can be a rate-limiting step in scale-up, design, and analysis.

In this talk, we present an integrated experimental and data analysis technique to dynamically characterize nanofiltration membranes, resulting in differential-algebraic process models with quantified uncertainty suitable for process design. Experiments were run in a 10 mL dead-end filtration Amicon 8010 stirred cell connected with a dialysate tank. We conducted transport studies by loading both the cell and tank with predetermined concentrations of the desired salt and pressurizing the system with nitrogen gas. This apparatus design allows for continuous monitoring of permeate sample mass. Both retentate and sample concentrations are measured at regular intervals. We then postulate a family of differential-algebraic equation models for the system. Nonlinear regression is used to successfully calibrate hydraulic permeability, solute permeability coefficient, and reflection coefficient parameters in the model. These parameters correspond to the membrane transport properties and depend on the membrane chemistry, membrane nanostructure, and thermodynamics of the membrane-solution interface. As proof of concept, we consider diafiltration experiments for Mg^{2+} and K^{+} ions across an NF-90 nanofiltration membrane. We cast nonlinear parameter estimation as a multi-objective least-squares optimization problem and explore the trade-offs between different experimental observations, i.e., the mass of collected samples and sample concentrations. Through this analysis, we determined reflection coefficient is not identifiable in conventional filtration experiments and used the model to inform the diafiltration experiment design that covers higher concentrations. Finally, we discuss how the proposed dynamic experiments and data analysis technique require much less data than classical steady-state experiments to create and validate a model.

References

Eugene, E., Phillip, W., and Dowling, A., Data science-enabled molecular-to-systems engineering for sustinable water treatment, *Current Opinion in Chemical Engineering*, **26**, 122-130, (2019a).

Eugene, E., Phillip, W., and Dowling, A., Material Property Goals to Enable Continuous Diafiltration Membrane Cascades for Lithium-ion Battery Recycling, *Computer Aided Chemical Engineering,* **47**, 469-474, (2019b).

Eugene, E., Phillip, W., and Dowling, A., Material Property Targets for Emerging Nanomaterials to Enable Point-of-Use and Point-of-Entry Water Treatment Systems, Preprint, https://doi.org/10.26434/chemrxiv.12526190, (2020).

Geise, G. M., Paul, D. R., Freeman, B. D., Fundamental water and salt transport properties of polymeric materials, *Progress in Polymer Science*, **39**(1), 1-42, (2014)

Hoffman, J. R., and Phillip, W. A., 100th Anniversary of Macromolecular Science Viewpoint: Integrated Membrane Systems, *ACS Macro Letters*, 9, 1267-1279, (2020).

Sadeghi, I., Kaner, P., and Asatekin, A., Controlling and Expanding the Selectivity of Filtration Membranes, *Mater.*, **30**(21), 7328–7354 (2018).

Yaroshchuk, A., Bruening, M. L., and Zholkovskiy, E., Modelling nanofiltration of electrolyte solutions, *Advances in Colloid and Interface Science*, **268**, 39-63, (2019).

619171 Simulation and Dynamic Analysis of an Integrated Plant for the Production of Methanol from Natural Gas

Abstract:

Methanol is an essential chemical for obtaining more complex chemical structures, such as acetic acid, methyl tertiary butyl ether, dimethyl ether, and methylamine. It is used in the manufacture of a variety of industrial and consumer products. In this research, a novel integrated plant for the production of methanol from natural gas is simulated and analyzed. The methanol manufacturing process is divided into three parts: (1) pre-treatment of natural gas, (2) conversion of natural gas to syn gas and (3) conversion of syngas to methanol. The pre-treatment step consists of compression and hydrodesulfurization stages. This is followed by steam reforming and secondary reforming to produce syn gas. The reformed gas is cooled and the excess heat is used for generating steam. The cooled syn gas is sent to a methanol synthesis reactor. Finally, a distillation train is used to purify methanol. The overall plant is simulated in the ASPENPlus environment and the predictive capabilities of this steady-state model is tested by comparing with plant data from a methanol synthesis plant in Beaumont, Texas. Then, a variety of process conditions are tested at steady state to optimize the production of methanol. Heat-integration tools are utilized for energy-saving and capital cost reduction opportunities. A comparative economic assessment based on existing plant information indicates that the use of process integration techniques has the potential to reduce costs significantly. Then, a dynamic model is developed in ASPEN Dynamics for the integrated plant and the effect of disturbances to the plant are tested. The dynamic model is utilized to conduct a variety of "what-if" studies to determine conditions where the production of methanol is optimized. Furthermore, opportunities for utilizing process control strategies to improve plant operation are studied.

619252 Machine-Learning Enabled Optimization of Force Fields

Abstract:

Molecular simulation is a powerful tool for studying the thermodynamic and dynamic properties of materials. Envisioning the future, molecular simulation shows great promise for screening vast molecular design spaces which could be expensive or infeasible to probe experimentally. The rise of molecular engineering, e.g., designer solvents, porous materials, energy storage technology, and biological systems, combined with ever-increasing computing power has enabled the field of computational molecular science and engineering to rapidly progress. With these new capabilities, simulations can be integrated within multiscale modeling and design frameworks and used to inform experimental design decisions. However, to utilize molecular simulation in this capacity requires quantitatively accurate molecular models, called force fields. Classical molecular simulations model intra- and intermolecular interactions with force fields, which use a functional form and parameters to describe the potential energy of a system. Developing generalized, or transferable, force fields to describe large swaths of chemical space has historically been a laborious endeavor, often taking months to years to complete. Though these off-the-shelf force fields offer accurate predictions for some systems, they inevitably lack quantitative accuracy across the extraordinary range of chemistries found in the natural and synthetic world. Further manual parameter tuning is often necessary to ensure the model has the required accuracy for the molecule(s) and properties of interest [Wang and Kollman, 2001]. Thus, force field optimization represents a bottleneck to applying molecular simulation to new systems.

In this work we propose a machine learning-enabled automated force field optimization framework. Specifically, we show integrating Gaussian Process (GP) regression (e.g., surrogate) models and support vector machine (SVM) classifiers facilitates rapid tuning of force fields and provides a quick and efficient route to highly accurate, physics-based molecular models. We train a GP surrogate model on the results of molecular simulations so that for a given set of parameters, our GP model predicts the resulting simulation experimental property prediction. We explore different methods of constructing our GP model, including various kernel and mean functions, to determine how to best harness a GP model's ability to identify optimal regions of force field model parameter space and obtain the most reliable simulation result predictions. We also examine the use of a SVM classifier to capture discontinuities within experimental properties which a GP model cannot replicate.

As a demonstration case, we optimize force fields for two hydrofluorocarbons (HFCs), HFC-32 and HFC-125, for properties including liquid and vapor densities, vapor pressure, and enthalpy of vaporization. Results show we can find at least 26 HFC-32 and 45 HFC-125 force field parameter sets in a timeframe of weeks which give mean absolute percent error in all of the properties of interest of at most 5%. Additionally, we find that these parameter sets are able to predict transport and critical properties accurately for HFC-32 and HFC-125 without the need for further tuning. We also applied our framework to tuning an ammonium perchlorate force field to predict solid properties, including lattice parameters, unit cell structure, and Hydrogen bond distances, angles, and symmetry. Multiple parameter sets have been found that outperform existing force fields in reproducing experimental observations of the listed quantities. Future work involves expanding this tuning method to encompass additional thermodynamic and transport properties of interest in an automated parameterization workflow. As further capabilities are developed, we envision this tool will be integrated within computer aided molecular and process design schemes to facilitate rapid multiscale design and optimization.

References:

Wang, J., & Kollman, P. A. (2001). Automatic parameterization of force field by systematic search and genetic algorithms. *Journal of Computational Chemistry*, *22*(12), 1219-1228.

619293 Thermodynamic Modeling of Aqueous Na^+-K^+-Ca^{2+}-Cl^--HCO_3^--CO_3^{2-}-CO_2 Solution

Abstract:

The enormous amount of highly saline water known as the "produced water" generated during oil and gas exploration can have salinity up to seven times that of seawater, raising a great concern for disposing it in the environment along with the scale deposition in the process equipment. To desalinate the water for further use, a comprehensive thermodynamic model is required. We are developing a comprehensive thermodynamic model to accurately calculate thermodynamic and phase equilibria properties, including salt precipitation, for high salinity produced waters [1]. As a part of the larger effort, the present study focuses on the development of a thermodynamic model for the aqueous electrolyte solutions involving sodium (Na^+), potassium (K^+), calcium (Ca^{2+}), chloride (Cl^-), carbonate (CO_3^{2-}), bicarbonate (HCO_3^-) ions, and molecular CO_2.

Utilizing the electrolyte nonrandom two-liquid (eNRTL) theory [2], this model requires parameterization of eNRTL binary interaction parameters and physical properties parameters of pure components, i.e., water, ions, molecular solute, and precipitating salts. With the fully parameterized eNRTL model, thermodynamic properties including CO_2 solubility and salt precipitation of the aqueous Na^+-K^+-Ca^{2+}-Cl^--HCO_3^--CO_3^{2-}-CO_2 solution and its subsystems can be accurately predicted. The model is validated against data at a temperature range of 273.15 – 473.15 K and electrolyte concentrations from infinitely dilution to salt saturation. The thermodynamic model developed is an indispensable tool to predict calcium carbonate scale deposition and support mass and energy balance calculations in desalination and water treatment processes. The model should also be an excellent tool for modeling CO_2 capture with brine solutions.

References:

[1] S. Tanveer and C.-C. Chen, "A comprehensive thermodynamic model for high salinity produced waters," *AIChE Journal,* vol. 66, no. 1, pp. 1-10, 2020.

[2] Y. Song and C.-C. Chen, "Symmetric electrolyte nonrandom two-liquid activity coefficient model," *Industrial & Engineering Chemistry Research,* vol. 48, no. 16, pp. 7788-7797, 2009.

AUTHOR INDEX

AIChE
120 Wall Street, FL 23
New York, NY 10005-4020

ISBN 978-1-7138-5340-4
9 781713 853404

CURRAN ASSOCIATES INC.
proceedings
.com

Management Division 2018

Core Programming Area at the 2018 AIChE Spring Meeting and
14th Global Congress on Process Safety

Orlando, Florida, USA
22 – 25 April 2018

AIChE
The Global Home of Chemical Engineers